Sustaining U.S. Nuclear Submarine Design Capabilities

Executive Summary

John F. Schank, Mark V. Arena, Paul DeLuca, Jessie Riposo
Kimberly Curry, Todd Weeks, James Chiesa

Prepared for the United States Navy

Approved for public release; distribution unlimited

The research described in this report was prepared for the United States Navy. The research was conducted in the RAND National Defense Research Institute, a federally funded research and development center sponsored by the Office of the Secretary of Defense, the Joint Staff, the Unified Combatant Commands, the Department of the Navy, the Marine Corps, the defense agencies, and the defense Intelligence Community under Contract W74V8H-06-C-0002.

Library of Congress Cataloging-in-Publication Data

Sustaining U.S. nuclear submarine design capabilities : executive summary /
John F. Schank ... [et al.].
p. cm.
ISBN 978-0-8330-4161-6 (pbk.)
1. Nuclear submarines—United States—Design and construction—21st century.
2. Shipbuilding industry—Employees—United States—21st century. 3. Navy-yards and naval stations—United States. I. Schank, John F. (John Frederic), 1946–

V858.S8712 2007
359.9'3—dc22

2007013350

Photo Courtesy of General Dynamics Electric Boat.

The RAND Corporation is a nonprofit research organization providing objective analysis and effective solutions that address the challenges facing the public and private sectors around the world. RAND's publications do not necessarily reflect the opinions of its research clients and sponsors.

Cover Design by Stephen Bloodsworth

Published 2007 by the RAND Corporation
1776 Main Street, P.O. Box 2138, Santa Monica, CA 90407-2138
1200 South Hayes Street, Arlington, VA 22202-5050
4570 Fifth Avenue, Suite 600, Pittsburgh, PA 15213-2665
RAND URL: http://www.rand.org/
To order RAND documents or to obtain additional information, contact
Distribution Services: Telephone: (310) 451-7002;
Fax: (310) 451-6915; Email: order@rand.org

Preface

For the first time, the U.S. Navy faces a period that could last a number of years in which there will be no design program under way for a new class of nuclear-powered submarine. The resulting lack of demand for the services of submarine designers and engineers raises concerns that this highly specialized capability could atrophy, burdening the next submarine design effort with extra costs, delays, and risks.

In 2005, the Program Executive Office (PEO) for Submarines asked the RAND Corporation to evaluate the cost and schedule impacts of various strategies for managing submarine design resources. Of concern were the design resources at Electric Boat and at Northrop Grumman Newport News, the two shipyards that have previously designed classes of nuclear submarines, as well as design resources at the key vendors that provide components for nuclear submarines. Also of concern were the technical resources of the various Navy organizations that oversee and participate in nuclear submarine design programs. RAND's analysis built on similar research RAND conducted for the United Kingdom's Ministry of Defence. This document summarizes the methods and findings of the research that RAND carried out for PEO Submarines.[1]

[1] For full documentation of this research, see John R. Schank, Mark V. Arena, Paul DeLuca, Jessie Riposo, Kimberly Curry, Todd Weeks, and James Chiesa, *Sustaining U.S. Nuclear Submarine Design Capabilities*, Santa Monica, Calif.: RAND Corporation, MG-608-NAVY, 2007. Available online at: http://www.rand.org/pubs/monographs/MG608/

This research was sponsored by the U.S. Navy and conducted within the Acquisition and Technology Policy Center of the RAND National Defense Research Institute, a federally funded research and development center sponsored by the Office of the Secretary of Defense, the Joint Staff, the Unified Combatant Commands, the Department of the Navy, the Marine Corps, the defense agencies, and the defense Intelligence Community.

The lead author of this report, John F. Schank, can be reached at schank@rand.org. For more information on RAND's Acquisition and Technology Policy Center, contact the Director, Philip Antón. He can be reached by email at atpc-director@rand.org; by phone at 310-393-0411, extension 7798; or by mail at the RAND Corporation, 1776 Main Street, Santa Monica, California 90407-2138. More information about RAND is available at www.rand.org.

Contents

Figures

Tables

Acknowledgments

This research could not have been accomplished without the assistance of many individuals. RADM William Hilarides, Program Executive Officer for Submarines, continually encouraged and supported the research effort. CAPT Dave Johnson, *Virginia* Class Program Manager, provided overall guidance to our efforts. John Leadmon, Scott McCain, and, from the Naval Sea Systems Command, CAPT Jeff Reed from the Ship Design, Integration and Engineering Directorate (SEA 05) and Carl Oosterman from the Nuclear Propulsion Directorate (SEA 08) graciously shared their time and expertise.

Numerous individuals at Electric Boat and Northrop Grumman Newport News shared their knowledge of the submarine design process and provided the data necessary to accomplish our analysis. We particularly thank John Casey, Steve Ruzzo, Ray Williams, Tom Plante, and Tod Schaefer from Electric Boat and Becky Stewart, Charlie Butler, and Don Hamadyk from Northrop Grumman Newport News.

We also appreciate the time provided by the numerous vendors that support nuclear submarine design efforts. Their responses to our surveys helped us understand the problems they face in sustaining their design resources. We would particularly like to thank Carol Armstrong of Northrop Grumman Sunnyvale.

Numerous individuals from the U.S. Navy provided insights into the roles and responsibilities of the Navy's technical community. Larry Tarasek and Daniel Dozier at the Naval Surface Warfare Center, Carderock Division, and Frank Molino at the Naval Undersea Warfare

Center, Newport Division, shared their extensive knowledge of their organizations' roles in nuclear submarine development.

Ron Fricker from the Naval Postgraduate School and Giles K. Smith of RAND offered valuable insights and suggestions on earlier drafts of this document that greatly improved the presentation of the research. At RAND, Robert Lien helped formulate the early analysis of viability of design resources at the suppliers. Deborah Peetz provided overall support to the research effort. John Birkler and Irv Blickstein shared their expertise and knowledge of industrial base issues, especially within the submarine industrial base.

The above-mentioned individuals, and others too numerous to mention, provided functional information, data, and comments during our study. The authors, however, are solely responsible for the interpretation of the information and data and the judgments and conclusions drawn. And, of course, we alone are responsible for any errors.

Abbreviations

CVN	carrier vessel, nuclear
EB	Electric Boat
HM&E	hull, mechanical, and electrical
MMP	multimission platform
NAVSEA	Naval Sea Systems Command
NGNN	Northrop Grumman Newport News
NSWC	Naval Surface Warfare Center
PEO	program executive office
SSBN	ship submersible ballistic, nuclear
SSGN	ship submersible guided, nuclear
SSN	ship submersible, nuclear
VLS	Vertical Launch System

CHAPTER ONE

Introduction

What Should Be Done About the Current Gap Between Submarine Design Efforts?

The U.S. submarine fleet currently numbers more than 50 fast attack submarines (SSNs) and 18 submarines built to launch ballistic missiles (SSBNs). All are nuclear powered to maximize the duration and speed of underwater operations. While the submarine fleet has been decreasing in size since the end of the Cold War, it is anticipated that the U.S. Navy will sustain a force of several dozen boats into the foreseeable future.

Submarines are almost continually being built to replace older ones that must be retired. As is the case with surface ships, submarines are built in classes—sets of boats constructed to a common design. Designing a new class of nuclear submarines is a very large and complex endeavor, lasting 15 years or longer and requiring 15,000 to 20,000 man-years at the prime shipyard contractor alone.

For the first time since the advent of the nuclear-powered submarine, no new submarine design is under way or about to get under way following the winding down of the current effort (for the *Virginia* class of SSNs, now in production). This is a matter of some concern: Submarine design requires skills developed over many years that are not readily exercised in other domains. The erosion of the submarine design base—at the shipyards, the suppliers to the shipyards, and the Navy itself—may lead to the loss of the required skills before a new design does get under way, perhaps in another six to eight years. This skill loss could result in schedule delays to allow for retraining, with consequent higher program costs and potential risks to system performance and safety. This raises the question of whether some action should be taken to sustain a portion of the design workforce over the gap in demand.

In view of these potential problems, we sought to answer the following questions:

- How much of the submarine design workforce at the shipyards would need to be sustained for the least costly transition to the next design? What are the implications of different approaches to allocating the workload?
- To what extent is the shipyard supplier base also at risk?
- How will the Navy's own design skills be affected by a gap, and how easily might they be recovered?
- Taking all answers to the preceding questions into account, what steps should the Navy take in the near future?

We take up each of these questions in turn in the subsequent chapters of this summary. However, by way of background, we first give a brief history of U.S. nuclear submarine design.

Motivators of New Submarine Design Have Evolved

The early years of nuclear submarine design were marked by experimentation. A new design was undertaken even before work had finished on the previous one, and few boats were built to the same design. As the Navy and the builders gained experience and winnowed the spectrum of alternative approaches to submarine design, some stability was achieved. The *Sturgeon* class, the first of which was commissioned in 1966, extended to 37 boats. Still, the evolution of the Soviet threat required the introduction of new designs in response. The *Los Angeles* class was introduced to service in 1976 and went through two additional "flights," or significant design updates (one of which included the incorporation of the Vertical Launch System [VLS] for cruise missiles), over the next 20 years. The *Seawolf* class was the last Cold War submarine class; production of this class was terminated after three boats in recognition of the end of that era. In the post–Cold War era, submarine design has reflected the changing threat: Some ballistic missile–carrying boats of the *Ohio* class have been partially redesigned to carry guided missiles instead (thus becoming SSGNs), and more attention is being paid to submarines' special-forces transport and support function. One of the *Seawolf*-class boats has accordingly been outfitted with a multimission platform (MMP) to allow for a more flexible interface with the ocean. Figure 1.1 shows the overlapping durations of submarine design efforts over the past 40 years, with bar colors indicating which shipyard, Electric Boat (EB) or Northrop Grumman Newport News (NGNN), performed the design effort.

Notwithstanding continued responsiveness to the evolving threat, new designs are now largely driven by the need to replace older boats that are wearing out. Currently, the already designed Virginia class is replacing all retiring boats. Thus, for the first time since the advent of

Figure 1.1
Overlapping U.S. Submarine Design Efforts Are Giving Way to a Gap in Demand

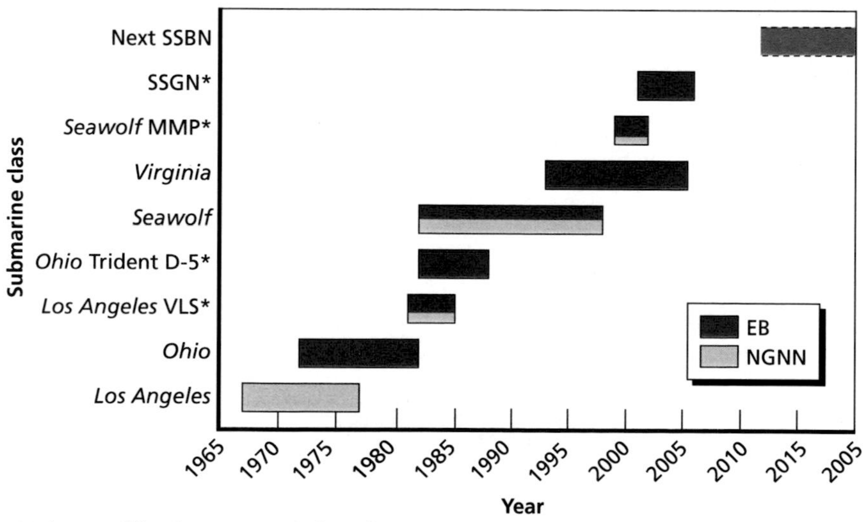

*Major modification to an existing class.

RAND *MG608/1-1.1*

nuclear power, no new submarine design is on the drawing board, and, according to current Navy plans, none will be until design work needs to get under way (perhaps sometime in the middle of the next decade) for a new SSBN class to replace the *Ohio* class.

CHAPTER TWO

Framing the Shipyard Analysis

To understand the results from our shipyard analysis, it is important to understand how we framed it. We broke the problem down into three parts: predicting design demand, formulating supply options, and estimating their costs. (We describe our approach to the supplier and Navy resources analyses in Chapters Five and Six.)

Step 1: Predict Design Demand

The first step in analyzing design workforce management options for the shipyards is to predict the demand for the next submarine design and its timing, beginning with the known demands—the design work "on the books." The latter involves both support to construction efforts on in-service submarines and to any new design efforts for surface ships, such as the CVN 78 class of aircraft carriers, or for major modifications to the *Virginia* class of SSNs. Then estimates are needed as to when a new design effort might begin, how long it would take, and the magnitude of the workload demand. Guided by the current 30-year shipbuilding plan and the prospective retirement dates of submarines in service, we infer that the next design effort will be for the next SSBN class, to begin in 2014. Assuming that the next design effort would be similar to that of the *Virginia* class, it will last 15 years and require approximately 35 million design and engineering man-hours. Because of the uncertainties inherent in such a projection, we examine the sensitivity of the cost and workforce management results to different start dates, durations, and workloads.

One virtue of the 2014 design start date is that the SSBN design effort would wind down in the 2020s, about the time the design of a replacement for the *Virginia* class will be ramping up (see Figure 2.1). Such a long-term view should be part of the submarine acquisition planning process, because a skilled workforce must be managed with the long view in mind. If the SSBN design were delayed by four or five years, it would overlap too much with the next SSN design, creating a longer near-term gap and a higher peak than shown in Figure 2.1. If it started much earlier than 2014 and still lasted 15 years, the current gap in demand could be replaced by one between the two peaks shown in the figure.

Step 2: Formulate Supply Options

Over the nearer term, given the anticipated SSBN design demand, how should the labor supply be managed? We categorize the available choices into two broad approaches—"doing nothing" and "doing something." Under the first approach, the prime contractors would adjust their workforce to meet demands only. This is shown in Figure 2.2, which schematically depicts the demand, along with a supply line intended to match it. The figure shows the future SSBN design demand on top of the ongoing design demand in the yard, e.g., to support the needs of submarines in service. The contractor allows the workforce to dissipate along with the demand for it and then builds the workforce back up when demand starts increasing. However, the new hires will not be as productive as the current workforce and, as a result, the work will take longer and cost more, as indicated by the yellow wedge in the figure.

In the "do something" option, the contractor would sustain a number of designers and engineers above demands during the gap to serve as a foundation to rebuild the workforce for a new design effort (see Figure 2.3). The productivity deficit for this rebuilt workforce would be less than that in the "do nothing" case, and there would be

Figure 2.1
SSBN Class Start Date Affects Design Demand Peaks and Gaps (case shown assumes 2014 start date)

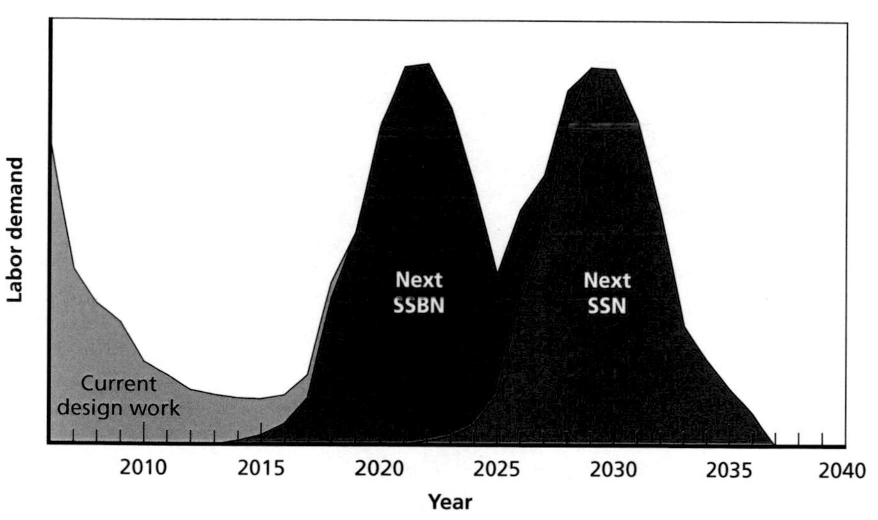

RAND *MG608/1-2.1*

Figure 2.2
The "Do Nothing" Option Leads to Long-Term Growth in Schedule and Workload

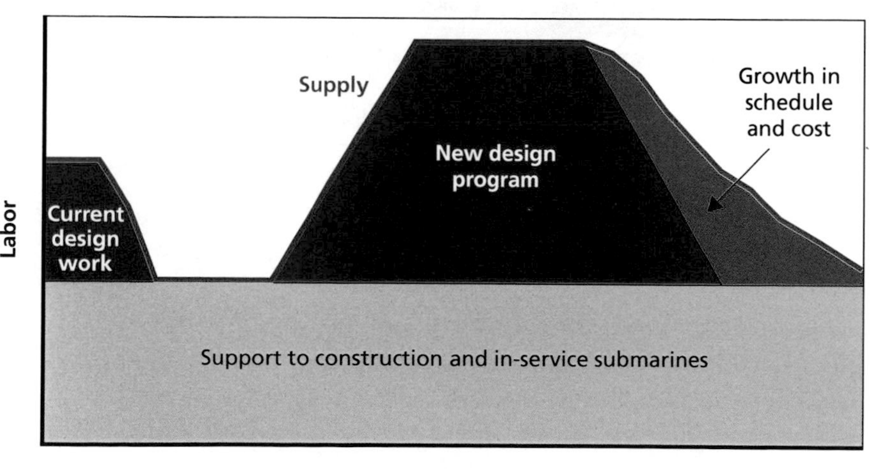

RAND *MG608/1-2.2*

Figure 2.3
The "Do Something" Option Trades Higher Near-Term Costs Against Long-Term Costs and Delays

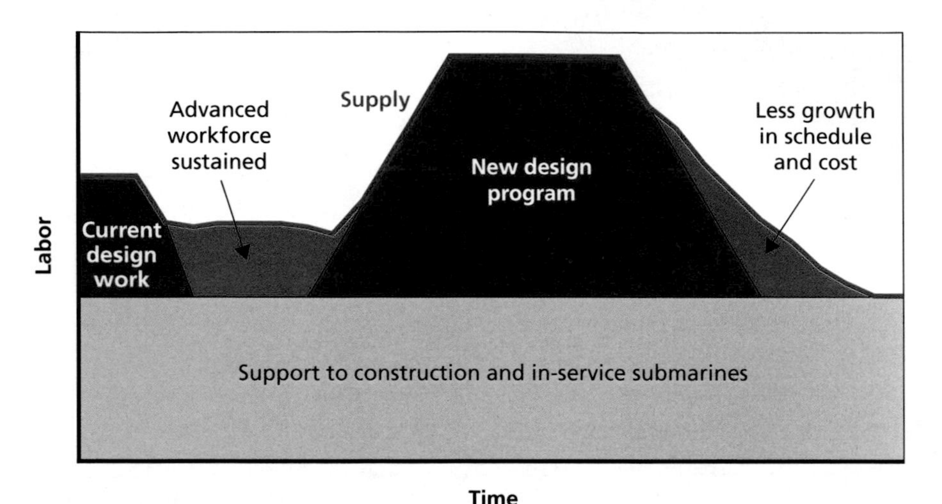

RAND *MG608/1-2.3*

a smaller eventual cost and delay penalty for not having the full workforce on hand when design ramps up. Whether this is cheaper than "doing nothing" depends on whether the money saved later is greater than the cost of sustaining design workers in excess of demand over the short run.

Step 3: Estimate Costs of Supply Options

Given these demand and supply relations, the next step in the analysis is to quantify the costs of the "do-nothing" and "do-something" strategies. Costs for different workforce drawdown and buildup profiles vary because of termination costs and hiring and training costs, as well as the efficiency-related penalties mentioned earlier. RAND has previously quantified the costs of production gaps; however, that research was focused on production workers. For design workers, we would expect, on the one hand, lower penalties from lost learning, because there is an inherent novelty to each succeeding design effort, but, on the other hand, higher penalties for the potential loss of expertise, which should take longer to accumulate for design than for production. Productivity losses, along with the costs of training, hiring, and termination, are estimated in a workforce simulation model that we developed. Both shipyards provided data for estimating these productivity losses and costs.

The model projects the workforce by skill category, age bracket, and experience level. It steps through time, adjusting the workforce according to the management option chosen ("do nothing" or "do something") and calculates the impact on the schedule and workload of a new design effort based on the composition of the design workforce when the new effort begins. The model computes total direct and indirect costs to compare the costs of sustaining various numbers of designers and engineers during the design gap.

The model then calculates the increase in schedule accruing from productivity losses and adjusts the workload upward to account for the fact that when prior submarine programs have experienced a given percentage increase in schedule, the result has been a similar percentage increase in cost. Workforce dynamics can result in other issues that the model does not consider, such as problems starting construction in time to meet a desired launch date because a delayed design process has not yet matured sufficiently. The model calculates the total cost of labor as equal to the sum of the costs associated with the design and engineering workforce over all time steps in the workforce simulation, plus the cost growth associated with the schedule penalty. In calculating the costs of reconstituting the design workforce after a gap, we take credit for retained designers and engineers as potential mentors; the more mentors, the faster the train-up for newly hired workers. However, we do not count as potential mentors that portion of the design workforce devoted to supporting construction or the operations and maintenance of in-service submarines. The application of submarine design skills in support is quite different from their application in new design.

The Results of the Analysis Should Be Interpreted with Caution

The following caveats apply to the results of our analysis:

- Our model does not produce budget-quality cost estimates. The results are best viewed as relative differences in the costs of alternative workforce management strategies rather than the absolute cost of any one strategy.
- All costs are estimates subject to estimating errors associated with future uncertainties. We do perform some sensitivity analysis on various workforce-related variables. Nonetheless, we cannot test for uncertainties for all conceivable parameters, so care should be taken in interpreting small differences in cost and other outcomes.
- Workforce-related model inputs are based on data received from EB and NGNN. We compare these data to similar values from other shipyards to ensure their reasonableness.
- We assume that both shipyards currently have the critical skills and proficiency necessary for submarine design. We do not test this assumption, which has implications for our results.

CHAPTER THREE

Workforce Levels and Costs for the Shipyards

What Size Design Workforce Is Least Costly for Different Yards and Workloads?

When we run the model, we find that, if the next SSBN is designed at EB, as has usually been the case in recent decades, and the "do nothing" approach is adopted, the design effort will take about three years longer than our nominal assumption of 15 years. Sustaining a workforce above the level needed to meet demand would cut back the increase in design duration. If 800 people could be sustained, there would be no increase. To the extent that there is insufficient demand to support them, the extra people cost money, but they also save money by precluding the extra work associated with the schedule delay and with workforce transition costs (termination, hiring, training). As shown in Figure 3.1, the net cost (relative to "doing nothing") decreases up to a sustained workforce of 800 people. Above that level, the up-front cost of adding each worker outweighs the increases in savings. Above a sustained workforce of about 1,400, the cumulative up-front cost exceeds the savings later on. Thus, the net cost is least when 800 people are sustained; that cost is about 10 percent less than what EB's "do nothing" approach would cost. Performing the same analysis for NGNN, the other submarine-capable design yard, indicates that 1,050 designers and engineers should be sustained and that doing so would save 36 percent of NGNN's "do nothing" design cost.

Figure 3.1
Base Case: At EB, Net Cost Is Lowest if 800 Workers Are Sustained

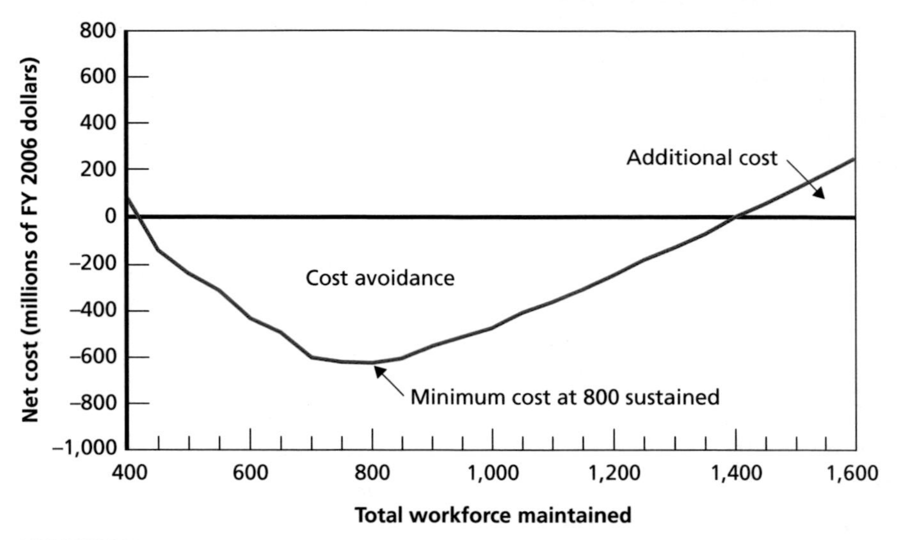

RAND *MG608/1-3.1*

The results given so far pertain to a start date of 2014 and a workload such as that required for the design of the *Virginia* class. The workforce to sustain (as indicated by where the curve bottoms out in Figure 3.1) is relatively insensitive to the start date but somewhat more sensitive to changes in the total workload (see Table 3.1). If the latter were to be 30 percent higher or lower than that for the *Virginia* class, the workforce to sustain would increase or decrease responsively—by 20 to 30 percent for most start dates at EB or NGNN. At the *Virginia*-class workload, however, the *total cost* would increase with later start dates (longer gaps) and decrease with earlier dates. The percentage saved relative to the "do nothing" approach would also be higher with higher workloads and later start dates and lower with lower workloads and earlier start dates. That is, at higher workloads, more would be saved relative to the cost of "doing nothing" at that higher workload than would be saved at lower workloads relative to the cost of "doing nothing" at that lower workload. At the expected 2014 start date, however, the sensitivity of percentage savings to workload would be small—covering a range of only 4 or 5 percentage points.

Table 3.1
Results for Different Design Workloads and Start Dates

	Results for Workloads Ranging from 30% Above to 30% Below *Virginia*-Class Design Workload, for Start Dates of		
	2009	**2014**	**2018**
EB			
Minimum-cost workforce to sustain	800–1,150	550–1,000	550–1,000
Percent labor cost savings relative to "doing nothing"	0–14	10–14	28–31
NGNN			
Minimum-cost workforce to sustain	850–1,400	700–1,200	700–1,200
Percent labor cost savings relative to "doing nothing"	2–17	37–42	41–46

NOTE: All savings are relative to doing nothing prior to the start date assumed and for the workload assumed.

Stretching the Work Results in Further Savings; Splitting the Work Does Not

So far, we have been assuming that an early start date would be followed by a 15-year design period. However, the design effort might be stretched to 20 years. This would have the benefit of filling the current design gap (and lowering the peak workload required; see Figure 3.2) without creating another gap once the SSBN effort is complete. The workforce sustained during the gap would be engaged in productive activity toward design of the new SSBN class. It would thus not be in excess of demand and would not be responsible for extra costs, potentially resulting in net savings.

Whether the design period is stretched or not, the Navy might consider it advantageous to split the design work between EB and NGNN, rather than retaining design expertise at only one firm. We examined both a 50/50 split and a 75/25 EB/NGNN split.

Figure 3.2
Stretching the Design Duration Can Fill the Gap

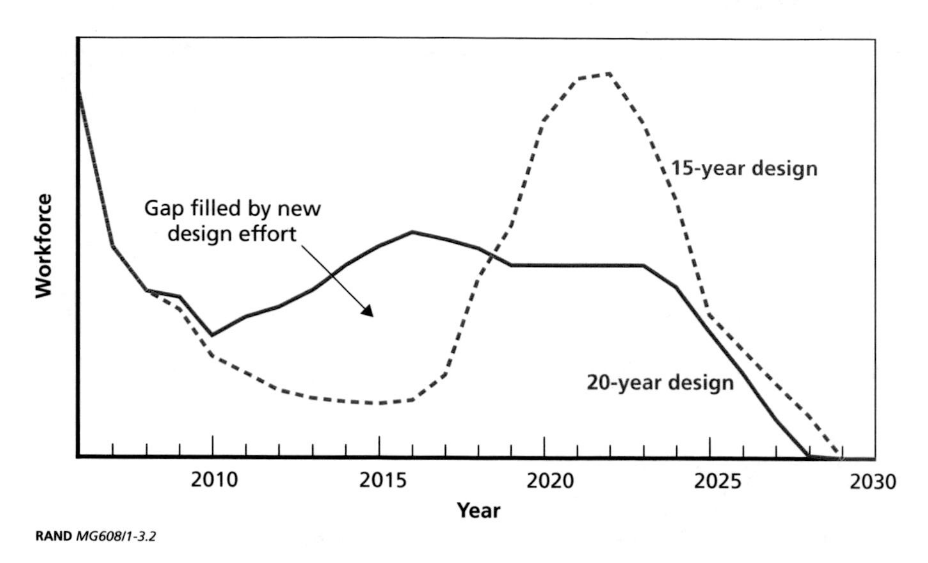

RAND *MG608/1-3.2*

The costs of these two approaches, singly and in combination, are shown in Table 3.2. If the work is not split, the 15-year design duration costs 21 percent more at EB than the 20-year duration, or, put another way, the 20-year duration costs 17 percent less. At NGNN, the 20-year duration costs 22 percent less. If the work is evenly split between EB and NGNN, the cost is somewhat higher than doing the work at one yard. This is true even without taking into account any inefficiencies involved in sharing the work. A 25 percent penalty for such inefficiencies might not be an unreasonable estimate, and the cost would increase accordingly. Costs for a 75/25 (EB/NGNN) split differ little from those for a 50/50 split.

While the 20-year approach looks promising, there are potential drawbacks. With the additional time, the Navy will have to expect and budget for additional iterations of technology refresh cycles. Additionally, the long design duration will increase opportunities to change requirements, which can also lead to increased costs. Finally, a stretched design duration could increase costs because the program effectively must pay fixed costs for an additional 4 to 5 years. However, the savings shown in Table 3.2 would help offset any fixed-cost penalty.

Table 3.2
Results for Different Workload Allocations

Yard	Design Duration (years)	Split (EB/NGNN)	Split Penalty (%)	Minimum Workforce Sustained[a]	Cost Increase over Minimum Cost Solution (%)[b]
EB alone	15	N/A	N/A	800	21
	20	N/A	N/A	900	0
NGNN alone	15	N/A	N/A	1,050	28
	20	N/A	N/A	950	0
Two yards (EB/NGNN shares)	15	50/50	0	950	31
			25	1,150	65
		75/25	0	900	31
			25	1,100	66
	20	50/50	0	975	17
			25	1,350	41
		75/25	0	950	14
			25	1,150	41

N/A = not applicable.

[a] For least costly workforce.

[b] The cost for the 20-year design profile at the shipyard.

How Sensitive Are the Results to Variations in Assumed Parameters?

Optimistic and pessimistic scenarios help to test how sensitive our results would be to variations in some of the parameters associated with the workforce: productivity, attrition, and hiring rate. In the optimistic scenario, the annual productivity gain and hiring rate are set 5 percentage points higher and annual attrition 1 percentage point lower. In the pessimistic scenario, the parameters vary by the same amounts in the opposite directions. These variations are consistent with those reported in the literature. Model runs indicate that, in these alternative scenarios, the minimum-cost workforce sustained would vary by 150 to 200 people—higher in the pessimistic scenario and lower in the optimistic one. Costs, of course, follow. At EB, costs in the optimistic scenario would be about 5 percent below those for the 15-year design baseline, and in the pessimistic scenario, more than 20 percent higher. Thus, reasonably more optimistic assumptions do not lead to significantly lower cost estimates, but reasonably more pessimistic ones can lead to somewhat higher cost estimates.

Sustaining the Skills of a Workforce in Excess of Demand Is Problematic

Sustaining a workforce in excess of demand is not without its drawbacks. Chief among them is the possibility that, in the absence of work on a new submarine class, skills might atrophy anyway, or skilled personnel might leave the workforce. Task options to mitigate this eventuality include

- spiral development of the *Virginia* class (i.e., technology or other capability upgrades over the course of production)
- design of conventional submarines, either for the U.S. Navy or for foreign sales
- conceptual design of new submarines with no intention to put them into production
- design of aircraft carriers or other ships
- collaboration with the United Kingdom or another allied government on one of its submarine design efforts.

These options all furnish some opportunity to maintain skills, but they all also have disadvantages. It is not clear that, even employed in combination or in coordination with other design activities, any or all of these would be able to sustain critical submarine design skills without some loss of capability. Of course, "doing nothing" will almost certainly result in substantial skill erosion.

CHAPTER FOUR

Critical Skills at the Shipyards

Hundreds of Technical Skills Are Required to Design a Submarine

A wide range of skills and technical competencies is required to successfully complete a submarine design. Recognizing that a gap in design efforts was imminent, EB undertook the categorization of the skills required for a submarine design effort. The result is the hierarchical paradigm shown in Figure 4.1. Using this paradigm and information from NGNN as a basis, we created 16 high-level skill categories.

Figure 4.1
Categorization of Nuclear Submarine Design Skills by Electric Boat

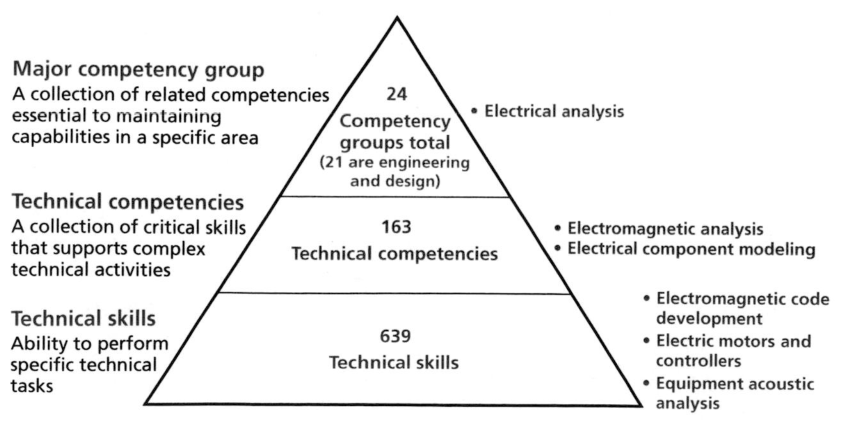

RAND *MG608/1-4.1*

The Skills of a Sustained Workforce Might Follow Their Distribution for the *Virginia*-Class Design

We have established the need to sustain 800 designers and engineers at EB, or 1,050 at NGNN, through the design gap if costs are to be minimized. These numbers should include representatives from all the skill categories, to ensure that all skills will survive a gap and that there will be an adequate base of mentors to reconstitute those skills in the workforce. For a first-order quantification of the number required in each category, we start with the allocation across categories for the *Virginia*-class design for EB, and, for NGNN, the allocation provided to us by the yard for a *Virginia*-size design effort. We apply that allocation to the 800 designers and engineers sustained at EB, or the 1,050 sustained at NGNN. The result is the distribution shown in Table 4.1.

But a Variety of Factors Should Be Taken into Account

Note that the preceding is only a rough estimate. The specific number needed to be sustained from each skill group will depend on various factors relating to the probability of losing and the difficulty of reconstituting each skill. These factors include

- The technical specifications of the next submarine design. If there is expected to be a significant change from the current design, the distribution of skills to retain should reflect that. For example, if it is likely that the next design will use electric drive, more electrical and fewer mechanical engineers will be required.
- Workforce demographics. Skill groups with older workforces need more management attention to ensure that a critical mass is not lost. About half the planning and production workforce at NGNN and most of the engineering support workforce at EB are over 50 years of age.
- Ability to find skills outside the nuclear submarine industry. Certain skills may be exercised in nuclear submarine design only, e.g., acoustics engineers and signals analysts who specialize in silencing

Table 4.1
Rough Estimate of Skilled Personnel to Sustain

	EB	NGNN
Designers		
Electrical	60	80
Mechanical	50	95
Piping/ventilation	65	95
Structural	80	95
Other	35	115
Designer subtotal	290	480
Engineers		
Electrical	40	40
Mechanical	65	60
Fluids	80	60
Naval architecture and structures	110	145
Combat systems	40	30
Acoustics	25	40
Planning/production	10	60
Testing	10	10
Management	10	105
Engineering support	50	10
Other engineering	80	10
Engineering subtotal	520	570
Total designers and engineers	810	1,050

and structural engineers specializing in shock. If these skills are lost, reconstituting them will be more challenging than for other types of skills.

- Time to gain proficiency. Skills that take a particularly long time to develop, because they require either a great deal of formal education or occupational training time, are also more challenging to reconstitute than skills that take less time to develop. Approximately 10 percent of technical skills, for example, require 10 years of on-the-job experience to develop—in some cases, following a Ph.D. (e.g., turbulence modeling, computational hull design and analysis, and nuclear containment analysis).
- Supply and demand factors. These may affect the availability of certain skills or the ease with which individuals with particular skills can be attracted to the industry. The number of nuclear engineering programs in U.S. universities, for example, has fallen by about half over the past 30 years. Partly as a result, the supply of workers is decreasing in certain key areas. At the same time, the U.S. Department of Energy forecasts that new nuclear power plants will be needed by 2025, which suggests a competing demand for nuclear engineers.

CHAPTER FIVE

Suppliers

Submarines, like other large, complex systems, are not designed by a single firm. A single firm cannot productively sustain all the special skills required. The submarine design base thus includes a large number of subcontractors that contribute design expertise or engineered components to plug into the system.

How Will Suppliers to the Shipyards Weather the Design Gap?

To find out, we surveyed suppliers identified by the shipbuilders as having significant activities associated with submarine design. We received responses from 38 of the 58 firms the shipbuilders identified; 32 felt that they had significant activities associated with submarine design. We analyzed these 32 responses according to a set of indicators of potential risk in the design industrial base:

- Percentage of revenue generated by design work. Only one firm got most of its revenue from design. Considered alone, this suggests that most firms could weather a design gap.
- Percentage of revenue from submarine business. Over three-quarters of the firms got half or less—usually much less—of their revenue from the submarine business—another indicator that a design gap would not have a large impact.

- Absence of competitors. Only five firms believed that they had no competitors, suggesting that in the event that some suppliers fail, the shipbuilder will typically have alternatives.
- Insufficient design workforce supply. Over a third of the suppliers indicated that they foresee a problem maintaining a technical workforce within the next 10 years—a period that extends through the expected SSBN design start date. About half foresee trouble beyond that.
- Percentage of workforce in upper age range. The average age of the design staff at over half of the firms is more than 45 years (see Figure 5.1). This is problematic because it suggests that many workers could approach retirement over the course of a submarine design gap. Such workers will not only be unavailable to meet workforce demand, they will not be there to mentor younger workers.
- Extent to which employment falls short of demand peak for design. Eighty percent or more of firms indicated that they already had sufficient staff to meet the peak design demand from a new submarine program.

Figure 5.1
At Most Firms, Most of the Design Staff Is Over 45

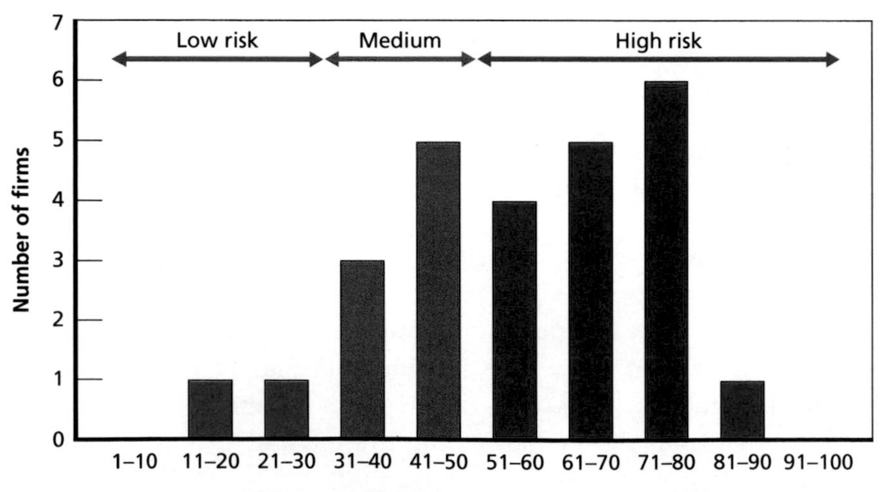

RAND *MG608/1-5.1*

- Time required to ramp up a design staff. Two-thirds of the firms thought that it would take a year or less to ramp up for a new submarine design effort. There should be sufficient notice to allow for that.
- Time required before a new hire is productive. Most respondents judged that it would take over six months for new hires to become adapted to the firm and proficient in their roles.

Some Suppliers Might Not Be Able to Offer Continued Support

The survey results suggest some reason for concern. While we cite favorable majorities for most individual risks, when we take them all together, over half of the responding firms (19 of the 32) show a degree of risk (see Figure 5.2). We judged eight of those firms to be at high risk because of risky scores in multiple dimensions.

Figure 5.2
Distribution of Vendors Across Risk Categories

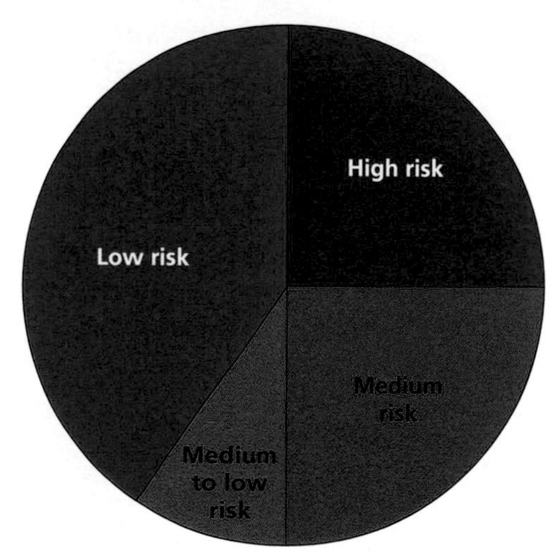

RAND *MG608/1-5.2*

Options Addressing Supplier Risk Need to Be Tailored to the Vendor

There are several possible options available for addressing supplier risk. Three of the options forgo the need for a new component design and two try to help vendors sustain and build a design staff during the gap.

- *Seek a competitive solution.* If the component or technology is not unique to a specific supplier, an alternative source could be sought. This option is feasible only if competitors exist and are able to maintain their design resources during the current design gap.
- *Replace the technology.* The next submarine class may not need a specific component that a vendor at risk provides. For example, it is possible that a new technology may replace an existing one.
- *Reuse the current component design.* A component may not need to be redesigned for the next submarine class if it can meet the needed performance attributes. Thus, design work could be avoided.
- *Stretch the next new submarine design period.* As with the shipbuilders, it might be possible to extend component design over a longer period of time to reduce the peak in the design workload.
- *Use spiral development for the* Virginia *class.* In an attempt to maintain design staffs, the Navy could initiate modernization design work with an at-risk supplier. This would work best for components in systems, such as combat or communication, that do not require significant layout or structural changes.

Most of these options are not applicable to all suppliers, as the situations of the different firms vary. In particular, stretching the design duration, a promising option for addressing the design gap at the shipyard, will not work for most of the vendors. The choice of intervention, or mix of interventions, will have to be tailored to each at-risk vendor.

CHAPTER SIX

Effect of a Design Gap on Navy Resources

The Navy retains ultimate responsibility for a safe, effective, and affordable submarine design. This responsibility has not changed despite significant changes in the division of labor between the Navy and private industry and in design tools and practices.

The Navy Holds Key Design Roles

In carrying out its responsibility, the Navy fulfills three roles: providing technical infrastructure and expertise, designing and developing certain critical components, and supporting submarine-related science and technology.

In providing technical infrastructure and expertise, the Navy plays the role of smart buyer. That is, it must ensure that the design efficiently meets Navy program requirements. In this capacity, for example, the Navy implemented integrated process and product development in the design of the *Virginia* class, an innovation intended to save time and money by making Navy design reviews a part of the ongoing effort, rather than a milestone occurrence. Another aspect of the infrastructure and expertise provided by the Navy is its role as a technical authority. This role is taken on specifically by an array of technical warrant holders, each of whom certifies within his or her area of expertise that the design is safe, technically feasible, and affordable. Finally, the Navy is responsible for design-phase testing and evaluation.

The Navy retains sole responsibility for designing and developing components that are associated with the nuclear propulsion plant, critical to submarine safety, critical to the integration and interoperability of the command-and-communication and combat-control systems, or not commercially viable for private industry to design. Submarine-related science and technology is integrated through the SUBTECH program, which consists of integrated product teams focusing on communications, weapon systems, self-defense, and hull and propulsion issues.

One of the strengths of the Navy's acquisition process is the separation of the responsibility for managing acquisition programs from the technical approval process. Program managers are responsible for program performance in cost and schedule terms. The Navy's technical establishment is responsible for the technical acceptability of the product design. In this way, safety issues are not subject to trade-offs against costs or schedule concerns.

Navy Design Activities Are Carried Out Mainly by the Naval Sea Systems Command and the Warfare Centers

The Navy's design resources are physically and organizationally dispersed between the headquarters of the Naval Sea Systems Command (NAVSEA) and its naval warfare centers. NAVSEA engineers oversee the design, construction, and support of the Navy's fleet of ships, submarines, and combat systems. The Naval Warfare Centers are charged with carrying out many of the specific activities supporting the Navy's design responsibilities. The Naval Surface Warfare Center (NSWC) is responsible for hull, mechanical, and electrical (HM&E) systems and propulsors for both surface and undersea vessels. The Naval Undersea Warfare Center is responsible for the bulk of submarine design issues.

The current division of responsibilities between NAVSEA and the warfare centers reflects a transition from a state in which more people were housed within NAVSEA. A major purpose of that transition was to move staffing from mission-funded positions, billable to Navy overhead, to program-funded positions, billable to a program executive office. The warfare centers operate somewhat like private contractors, billing their time to specific accounts and moving personnel to wherever the work is needed. This has implications for the conservation of submarine design expertise in the Navy.

As with the shipyards, a design gap could affect the Navy through personnel termination, consequent skill loss, impediments to the development of managers, and eventual hiring and training or rehiring and retraining, with all the costs those involve. There is also the possibility that some skills, once lost, could be difficult to reconstitute.

NAVSEA Would Not Lose Personnel but Could Lose Some Expertise

The specific effects of a gap would vary by organization. As a mission-funded organization, NAVSEA's technical infrastructure would likely survive a submarine design gap. However, the lack of ongoing design programs could degrade NAVSEA's ability to properly develop ship design managers. In particular, the lack of an ongoing new submarine design effort will mean that these engineers will not have an opportunity to exercise their whole-ship integration skills. The gap will also retard the development of senior managers capable of providing leadership during subsequent design efforts. Finally, proficiency in creating detailed technical specifications will decrease in the absence of a design program.

The Warfare Centers Need at Least $30 Million per Year to Keep from Losing Skilled Design Professionals

The impact of a design gap on the naval warfare centers depends on the technical areas involved. Non-HM&E areas are relatively insensitive to the gap because work in these areas is performed at the Naval Undersea Warfare Center, where in-service modernization programs make up the bulk of program funding and provide a healthy technical basis for new submarine design. However, at the NSWC's Carderock Division, ongoing in-service submarine support, technical assistance to the *Virginia*-class production program, and science and technology programs will not sustain the skills required for a full submarine design effort. As a result, engineers and designers who have been working on the *Virginia* design will shift to funded programs, i.e., those unrelated to submarines, or leave. Meanwhile, underused facilities (see Figure 6.1) might have to be laid up or placed on overhead accounts.

Carderock estimates the minimum workforce for sustaining design capability at 170, or about half that required for a full design effort (see Table 6.1). Carderock has received an average of $113 million per year in support of its submarine technology programs since the end of the Cold War. It would thus take about $55 million per year to support half the workforce and an equivalent proportion of the facilities. In-service support and technology development programs have averaged $23 million per year in funding. This leaves Carderock facing a $30 million to 35 million per year shortfall in the funding required to support its core technical group of personnel and facilities.

Figure 6.1
Most Prominent Design Facilities at NSWC's Carderock Division Are Underused

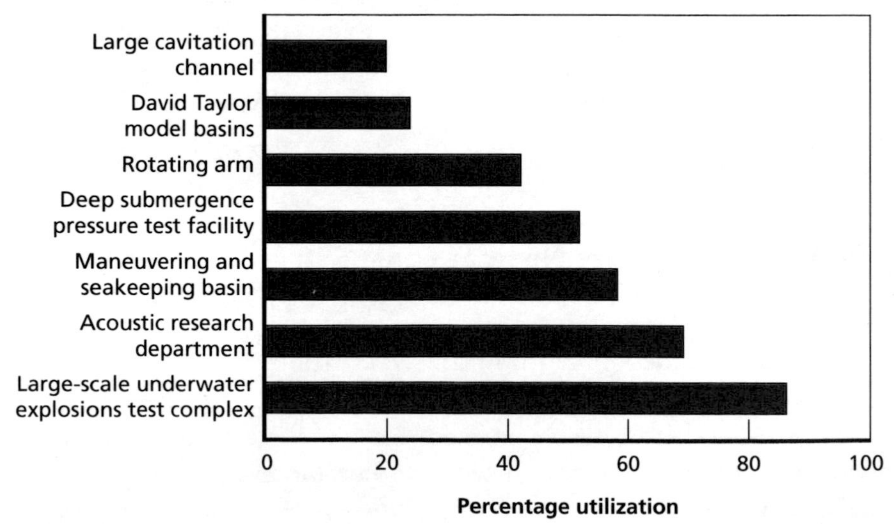

RAND *MG608/1-6.1*

This shortfall would be made up if the design duration were stretched to 20 years, essentially level-loading the Carderock division at the 170-person core design complement.

Table 6.1
Manning Levels to Sustain Design Capability and to Support a Full Submarine Design Effort at NSWC's Carderock Division

Technical Capability	Minimum Personnel Requiremed to Sustain Design Capability	Personnel Required to Support Full Submarine Design
Ship design and integration	4	14
Ship acquisition engineering	1	3
Hull forms, propulsors, and fluid mechanics	48	73
Mechanical power and propulsion systems	4	14
Electrical power and propulsion systems	4	10
Auxiliary machinery	7	22
Undersea vehicle sail and deployed systems	3	5
Surface, undersea, and weapon vehicle materials	10	15
Surface and undersea vehicle structures	11	15
Alternate energy and power sources research and development	1	2
Vehicle vulnerability, survivability, and force protection	14	20
Active and passive acoustic signatures and silencing systems	22	60
Nonacoustic signatures and silencing	5	17
Facility operations	36	79
Totals	170	349

CHAPTER SEVEN

Conclusions and Recommendations

The motivating concern for this research has been the potential for the loss of U.S. submarine design capability, given the gap in design demand inherent in the Navy's current shipbuilding plans. There are two aspects to this loss in capability—the loss of workforce capacity and the loss of critical skills. We have assessed the potential of both to erode capability at the shipyards, at the suppliers, and within the Navy itself.

We evaluated two basic workforce management strategies: (1) let the workforce erode and then rebuild it to design the next class of submarines and (2) sustain some number of workers in excess of those needed to meet the residual design demand during the gap. We found the latter to be less expensive. The number of workers to sustain depends on various assumptions. Consider a design duration similar to those for preceding classes (15 years), a workload similar to that for the *Virginia* class, and a start date for designing the next class that is consistent with current Navy ship replacement plans (2014). In that case, EB would accomplish the next design least expensively if, during the gap, it sustained a minimum of 800 designers and engineers, and NGNN if it sustained 1,050 (including those needed to meet the residual demand). These numbers vary up or down by a few hundred if workload and start date are varied over their likely ranges.

The design workload could also be varied both spatially and temporally. It could be split between the two shipyards, in an effort to maintain two capabilities. This does not convey an advantage in cost or in workforce sustained, even if it is assumed that division of the workload would cause no inefficiencies, which seems unlikely. The workload could also be stretched out over time. For example, the 15-year effort could be stretched to 20 years and, importantly, started early, in 2009, thus preempting most of the workforce drawdown. In that event, no extra workforce need be sustained to minimize cost (assuming that all the work is done by one yard), and the cost minimum would be lower than that achievable with a 15-year design. There are some drawbacks to stretching out the design, e.g., the greater possibility of design obsolescence by the time the first of class is launched, and these must be considered in any decision regarding this option. However, there is also an important drawback to sustaining workers in excess of demand: the need to find them something to do that will allow them to maintain those skills. Several options are available, but even in combination, these may not be sufficient for skill retention equivalent to that achievable by work on a new submarine class.

EB is addressing the specifics of the critical-skills problem, so we do not repeat that effort. However, we break out the recommended sustained workforces by general skill categories, based on information from the shipyards regarding the breakdown of the entire design workforce. We also offer some aggregate-level observations regarding the effect of the evolution of such skills on decisions as to which to support. We identify workforce demographics, time required to gain proficiency, and supply and demand as among the factors that should be considered.

The potential problems arising from a design gap extend beyond the shipyards. Numerous submarine components are provided by vendors that must design their products. We conducted a survey that asked firms about some of the issues common to critical shipyard skills (demographics, time to proficiency), as well as issues more specific to vendors (presence of competitors, percentage of work devoted to design). We found that, while in any one dimension most firms appeared likely not to encounter problems that would hinder their contribution to submarine design after a gap, some appeared to be potentially at risk in more than one dimension.

The Navy's roles in submarine design include exercising responsibility for ensuring that various aspects of design are consistent with safety and performance standards and designing certain components. We reviewed these roles, along with workforce structure and trends in pertinent Navy organizations, and came to the following conclusion: Sufficient design expertise in the various major skill categories was unlikely to be sustained to support HM&E submarine design functions at the NSWC's Carderock Division. Between $30 and 35 million per year would be required to sustain sufficient staff in submarine design in excess of those needed during the design gap. For both the Navy and for some vendors, avoiding the greater part of the design gap (e.g., by stretching out the design of the next class and starting it early) would obviate the need for concern over skill loss.

From the preceding analysis, we reach the following recommendations:

- Seriously consider starting the design of the next submarine class by 2009, to run 20 years, taking into account the substantial advantages and disadvantages involved.

If the 20-year-design alternative survives further evaluation, the issue of a gap in submarine design is resolved, and no further actions need be taken. If that alternative is judged too risky, we recommend the following:

- Thoroughly and critically evaluate the degree to which options such as the spiral development of the *Virginia* class or design without construction will be able to substitute for new-submarine design in allowing design professionals to retain their skills.

If options to sustain design personnel in excess of demand are judged on balance to offer clear advantages over letting the workforce erode, then the Navy should take the following actions:

- Request sufficient funding to sustain excess shipyard design workforces large enough to permit substantial savings in time and money later.

- Taking into account trends affecting the evolution of critical skills, continue efforts to determine which shipyard skills need action to preserve them within the sustained design core.
- Conduct a comprehensive analysis of vendors at high risk to determine the interventions required to preserve critical skills.
- Invest $30 million to 35 million annually in the NSWC's Carderock Division submarine-design workforce in excess of reimbursable demand to sustain skills that might otherwise be lost.